William Chimmo

Bed of the Atlantic

From One Sounding of 12,000 Feet Deep in the Atlantic Ocean

William Chimmo

Bed of the Atlantic
From One Sounding of 12,000 Feet Deep in the Atlantic Ocean

ISBN/EAN: 9783744691956

Printed in Europe, USA, Canada, Australia, Japan

Cover: Foto ©Andreas Hilbeck / pixelio.de

More available books at **www.hansebooks.com**

BED OF THE ATLANTIC.

FROM

ONE SOUNDING OF 12,000 FEET DEEP

IN

THE ATLANTIC OCEAN,

IN LATITUDE 47° NORTH, LONGITUDE 23° WEST,

ARE TAKEN UPWARDS OF

One Hundred (Microscopic Drawings of) Minute Organisms;

BEAUTIFULLY ILLUSTRATING

"In that great and wide Sea are things creeping innumerable."

BY

WILLIAM CHIMMO,

COMMANDER ROYAL NAVY;

FELLOW OF THE ROYAL GEOGRAPHICAL, ASTRONOMICAL, AND METEOROLOGICAL SOCIETIES OF LONDON.

[1870]

TO

ADMIRAL SIR JAMES HOPE, G.C.B.

ETC., ETC., ETC.,

IN GRATEFUL REMEMBRANCE OF
MANY KIND ACTS, BOTH PUBLIC AND PRIVATE, WHILE SERVING
UNDER HIS COMMAND IN THE WEST INDIES,

THIS LITTLE WORK

IS

DEDICATED

BY

THE AUTHOR.

PREFACE.

ONE year elapsed after these drawings had been completed before it was decided that they should be published.

This decision was greatly influenced by the encouraging remarks of one of the most skilled naturalists and microscopists in Ireland ; part of whose letter is here quoted, as an appropriate Introduction to this little work :—

" The interest that attaches to the subject of minute organisms at great depths, justifies the publication of any contributions on the subject, however small. Your labours have not been small. Whether you may or may not have discovered forms new to science, is a matter of little consideration : a new field is opened up to research, and it is something to stimulate enquiry. This is the least good likely to result from your publication. As to Diatoms, you have brought up *three* forms not previously described, and if to other departments your success has been equal to this, your labour has not been in vain.

" ' Macta puer virtute.'

" Let no man deprive you of the honour you have won, but cultivate assiduously the field you yourself have opened, and you shall have your reward."

WEYMOUTH, DORSET,
1870.

BED OF THE ATLANTIC.

A PAPER* was read, on the 8th of February, 1869, before the Royal Geographical Society of London, in connection with the Gulf Stream, being an investigation of its northern limits, its currents, depths, and temperatures; and in continuation of that paper, the present, which is one of natural history rather than geography, shewing a few of the beautiful organic remains found in those depths—some already known, but from their fantastic and elegant outline are again figured; others new, and attempted here to be drawn and briefly described: but it is feared that no adequate idea of their exquisite beauty and structure can be given with pen or pencil; while others, not altogether new, are doubtless varieties of those species well known. These, brought from depths of nearly three miles, under the influence of the warm waters of the Gulf Stream, whose development by temperature is so wonderfully evident, shews, in every minute form, unmistakeable evidences of design, which excite both admiration and wonder!

These are the productions of *one single* sounding from 2000 fathoms, and are upwards of one hundred in number!

(See Plates 1 to 14.)

A few remarks on some forms of Foraminiferæ, Polycystinæ, and Diatomaceæ, brought up from great depths under the Gulf Stream, between the Newfoundland Banks and the Azores—the parallel of latitude being 47° north, the temperature of the sea surface 70°, that of the bottom, at 2000 fathoms, 40° (Fahr.)

The naturalist will doubtless admit that it involves not only the expenditure of much time, but great labour to properly examine material collected from great ocean depths.

* See paper in Appendix, Geo. Soc., 8th February, 1869.

Therefore, the object of this paper is to give *simply* a sketch or outline of those minute and beautiful microscopic creatures brought up from a depth of 2000 fathoms (12,000 feet), under the influence of the Gulf Stream, and from which locality no specimens have as yet been collected, excepting those five hundred miles north on the line of the Atlantic cable, for the most part under the influence of Polar waters.

These few illustrations must be considered simply introductory: it is not easy for one not a naturalist to give every detail, but there is little difficulty in describing in a general manner what he sees.

A voyage through the Gulf Stream (by orders of the Lords Commissioners of the Admiralty) for the purpose of defining its limits, depths, currents, and temperatures, gave the writer an opportunity of sounding and bringing up the bottom, between the Newfoundland Banks and the Azores—a portion of the Atlantic which had not been before examined, and where it was supposed many banks and volcanic rocks existed.

The bottom has been found, for 60,000 square miles, to be a comparatively level chalk bed, made up of all those minute calcareous and silicious creatures found in the chalk cliffs of our own shores, chiefly Infusoria—in general, Foraminiferæ, in all its varied organic forms; frustules of Diatomaceæ, and Sponge Spicules, surrounded and closely packed by minute granules of inorganic silicious forms.

Around the great Banks of Newfoundland, and in those places under the influence of cool Polar waters, the colour of the ground was for the most part dark, brown, or green, and contained a greater number of silicious forms; while in the deeper waters, near the axis of the Gulf Stream, and where land was departed from, the colour changed to a pale yellow tint, and contained Foraminiferæ, for the most part, with few Diatoms, illustrating, in a very practical manner, the absence of plants.

On the sloping Banks of Newfoundland, which incline in a remarkably regular manner, from a depth of 1000 fathoms to 30 fathoms, at an angle of about 35°, and before leaving the shoal waters (at 600 fathoms), where the temperature was 38°, the bottom at that comparatively shoal depth may be said to be vegetative, with Sponge Spicules, fine silicious granules, and fragments of 'Coscinodiscus.

While in the deeper waters, and the level bottom of the Atlantic, calcareous shells of various forms of Foraminiferæ, silicious discs of the Polycystina—a variety of Coscinodiscus and Cocconeis—was found, and it is not improbable that here the lowest orders of animal forms are represented by the Foraminiferæ and Polycystina, and that the Coscinodiscus and Cocconeis represent those of plants.

At so great an abyssal depth of nearly three miles it is a subject of wonder how these masses of Foraminiferæ exist, under the many difficulties of absence of light, heat, air, &c., and subject to the enormous pressure of nearly $2\frac{1}{4}$ tons on the square inch. We have yet to learn on what these creatures exist!

From the greater depths, 2000 fathoms, were brought up skeletons of carbonate of lime and of silex; many of the latter contained Diatomaceæ in an unbroken and very perfect state, chiefly—

> Coscinodiscus, of various species;
> Nitzschia;
> Epithemia;
> Synedra, of various forms;
> Xanthiopyxis;
> Asterolampra;

with other forms not described before, but here figured.

The Euodia Chimmoensis, a new species, which has been fully described in the "Proceedings of the Dublin Microscopic Society;" but one species was before found by Bailley.

Several of these simple and delicate forms were found, some very minute, scarcely the ·002 (two-thousandth) of an inch; indeed, the more minute the form has been, the more calculated I found it to excite both admiration and wonder.

It was a singular fact, and one of some interest, that among these forms a large number belonged to the Diatomaceæ, when in the soundings of 1856, a few miles north of this position, it was found that "while Coscinodiscus abounded, frustules of ordinary Diatoms were exceedingly rare, and generally empty and broken." It will be seen that it is quite the reverse in this instance, all being in a capital state of preservation.

Among the Foraminiferæ were noticed abundance of the Globigerinæ, Orbulina,

B

and Rotalia, the universal characteristic, with Lagena, Lituola, and numerous Coccoliths and Coccospheres!—and many other varieties of forms not here figured.

Those from lesser depths present a marked contrast to the above; they are of a greater specific gravity, consisting chiefly of coarse grains of silex with Diatoms, but few Foraminiferæ, and those chiefly Globigerinæ and Coccoliths, for the most part water worn.

Many Polycystina, of various beautiful and fantastic forms, such as Podocyrtis, Astromma, Eucyrtidium, Lychnocanium, and numerous others, were also found, and some are here attempted to be figured, but no adequate description can be given, or idea formed, with pen or pencil, of such varied and exquisite forms.

Some of these illustrations are selections from drawings which were made on the first arrival of H.M.S. *Gannet* from her cruize in the Gulf Stream; but as those were intended only to give a very superficial and general idea of the nature of the bottom at so great a depth, the present plates have been substituted. More time and care have been devoted to them—many have been subjected to the full powers of one of Smith and Beck's best binocular microscopes—many forms magnified with ¼ and ⅛-powers, with shifting eye-pieces, to 300 and 850 diameters with the ¼-power, and to 900 and 1800 diameters with the ⅛-power.

To Dr. Moorhead, of Weymouth, whose beautiful microscope first revealed these interesting forms to me, and enabled me to make my first series of drawings, magnifying 252 specimens, I am very much indebted.

And to George Bishop, Esq., of Twickenham, whose powerful microscope shewed all the minute detail seen in the several series of sketches, I am particularly indebted; and if these few leaves should ever appear in print, I avail myself of the opportunity of recording my best thanks.

To the Rev. Eugene O'Meara, of Dublin, author of many interesting works on Diatomaceæ, who, from the beginning, took great interest in these deep-sea creatures, I owe many thanks for having corrected much of the orthography in these plates.

Many instructive ideas suggest themselves from these gleanings in the Gulf Stream—this Mighty Ocean River!—and the soundings collected from beneath it. These latter are worthy of much thought and investigation.

Thousands of these particles which appear under the microscope, are perfect organisms, formed by no other means than vitality itself; others, fragments of the same, either calcareous (composed of carbonate of lime collected by the animals themselves from ocean water) or silicious deposits from marine plants. All these Foraminiferæ were taken off the bottom (2000 fathoms), where it is supposed they lived and carried on their industrious and extensive work of chalk making (similar to the Polypii, the builders of those mighty coral walls and reefs) below water, independently of all those elements necessary for life above water.

It appears also not improbable that they live at different depths in the ocean —held in suspension, swimming, or otherwise moving, and for which the numerous pseudopodial arms of the Foraminiferæ appear well adapted.

It is not very long since it was supposed that many isolated dangers and irregularities—such as mud-banks, volcanic islands, rocks, and other vigias of vast extent—were dotted over the Atlantic Ocean. From these soundings (with others of former years) may be gathered a pretty correct and general idea of the arrangement of the Atlantic plateau.

From the west coast of the British Isles—France and Spain—to the east of North America and Newfoundland—from Greenland in the north, to the Azores in the south—may now be considered a comparatively level bed of chalk formation, caused by countless thousands of these invisible creatures (some here illustrated) before the world was. To what depth we cannot very accurately say, nor can we well imagine the rate of accumulation ; slowly and imperceptibly, no doubt. But this vast level plain, of upwards of 60,000 square miles, has scarcely an undulation (with the exception of those banks produced by the soil brought from the north by icebergs and deposited on the Newfoundland banks), and on whose bed safely and softly reposing, bathed also by the warm waters of the Gulf Stream, are our telegraphic cables, conveying our thoughts and wishes in an instant from the Old to the New World.

Notes.—It will be here seen that the *Scale* of many of these drawings has not been noted. To the naturalist this will probably appear an omission ; but they have been disregarded simply to give more scope to every detail of the forms, and have therefore been enlarged or otherwise at the will and pleasure of the artist.

A naturalist will at once detect the want of arrangement and classification, as well as the absence of artistic delineation in these interesting and beautiful forms, some of which I believe to be new, or at least varieties of species already known. This irregularity will, I am sure, be excused ; and it is not improbable that many have been overlooked by an unskilled microscopist.

These drawings were exhibited at the Soirée of the Royal Society, Burlington House, June, 1869.

ARRANGEMENT OF THE PLATES.

Diatomaceæ—Plate 1, 2, 3, 5, 7.
Polycystinæ—Plate 6, 11, 12, 13.
Foraminiferæ—Plate 4, 8, 9, 10.
Silicious Spicules—Plate 14.

Diatomaceæ (Silicious)—
  Coscinodiscus Lineatus.
    C. Crassus.
    C. Eccentricus.
  Cocconeis.
  Nitzschia Insignis.
  Epithemia Marina.
  Synedra, 3.
  Xanthiopyxis.
  Asteromphalus.
  Euodia (Chimmoensis), 4.
  Asterolampra.
  Coscinodiscus, Minor.
  Aulacodiscus, 2.
  Triceratium.
  Cymbella.
  Campylodiscus.
  Actinoptychus.
  Tryblionella ?
  Navicula.
  Licmophora.

Polycystinæ—
  Radiolaria.
  Eucyrtidium Elegans.
  Lychnocanium Falciferum.
  Astromma.

Polycystinæ (continued)—
Podocyrtis Mitra.
  „ Schomburgki.
Lychnocanium Lucerna.
Haliomma.
Eucyrtidium Tubulus.
E. Dyctyocha.

Infusoria, Foraminiferæ—
Globigerinæ.
Rotalia.
Lagena Vulgaris.
Polystomella.
Polymorphina Acuminata.
Orbulina.
Crystellaria.
Nonionina.
Spiroloculina (Miliolene 2).
Nodosaria, Radicula.
Lituola.
Entosolenia, Marginata.
  „ Globosa.
Bulimina, Elegantissima.
  „ Pupoides.
Dentalina, Subarcuata.
  „ Legumen.
Textularia, Variabilis.
Miliolina.
  „ Bicornis.
Spirilini.
Triloculina.
Biloculina, Ringens.
  Three forms not named.

15

Asterolampra (Bishopii).—This beautiful form, to which it is impossible to do justice in a drawing, is composed of seven radiates ; the cellular structure is absent from the rays except in the centre, where there are a few circular cells, largest in the centre ; margin smooth ; the outer marginal cells are of a deep brown tint and also cellular, which shewed the transparent form and radiations of the star very beautifully.

PLATE 1.

Fig. 1.

2000 fathoms
Lat. 47ᵛ N.
Long. 23° W.

The drawing may appear stiff and harsh, but it is an exact representation of this interesting species of Asterolampra. I have named this after my friend George Bishop, Esq., to perpetuate my sense of obligation to him.*

Triceratium (× 600).—Form triangular ; cells circular, eccentrically arranged ; sides concave ; angles neither elevated nor produced, but obtuse or rounded ; margin clear.

Fig. 2.

2000 fathoms
Lat. 47° N.
Long. 23° W.

The forms already known have generally convex sides, acute angles, elevated, cells regularly arranged and extending to the margin, and have been found for the most part in fossil remains. This is probably a variety, differing from those already described by the concavity of its sides, &c., and but one form was detected in all the soundings brought up.

Entosolenia (?)

These are separate drawings of different specimens (no doubt of the same genus). They are of a pinkish hue, with a clear white canal passing longitudinally and diagonally across the interior ; the lower part of fig. A was dark and spotted ; segments elongated ; arcuate tubes ; porcelainous, opaque, and glossy.

Figs. 3 & 4.

---

* This drawing was transferred to ivory by means of the ornamental turning-lathe, and, without altering or neglecting any of its detail, formed a very handsome and interesting model. Diameter, 2½ inches.

PLATE 1. *Miliolina* (B).

Fig. 3. *Lagena-form.*—Cell calcareous; form oval; neck with orifice passing along the external form; pseudopodial pits quite distinct; Miliolina Bicornis.

Fig. 2. *Spirilini Cornuspiræ.*—A simple spiral shell without chambers; tube undivided; convolutions in contact; aperture, diameter of tube, and simple; crest or crown of shell striated.

Diatomaceæ.

PLATE 2. *Coscinodiscii* (× 600; diameter .003 inch).—The forms are both
Fig. 1. an interesting and beautiful group; the most common are C eccentricus
and C minor, of which latter were in great numbers. The beautiful
2000 fathoms disc, here figured as No. 1, shews that each frustule is distinctly and
Lat. 47° N.
Long. 23° W. regularly formed of sexagonal shape. In this particular species the
areolæ in the centre are largest, decreasing towards the edge, the
margin finished by a double row of cells set vertically; the lines formed
by the cells are both radiating and eccentric, resembling the ornamental engine turning on the back of a watch.*

Fig. 2. *Actinoptychus* (.003 about).—Valves convex; cells circular; larger
2000 fathoms about the centre and margin; margin spinous; valve with six rays; it
is not improbable that this is the form above named, although the
side view A hardly gives the undulating form which this species
possesses.

* The hoops or zones which united the valves of the Coscinodiscus were found in
great numbers, broken in irregular fragments; but the number of valves found did not
at all coincide with the zones which united them.

Coscinodiscus.—Cells sexagonal; regularly arranged; largest in centre;   Fig. 3.
margin ornamented with double row of vertical cells.   2000 fathoms

Note.—Some of these forms may be Eupodiscii, although the elevated
processes were not visible.

These discoidal forms—Coscinodiscus, Eupodiscus, Asterolampra,
Asteromphalus, Aulacodiscus, &c., and many others—so much resemble
one another when subject only to a medium power, that an error in
classification will be excusable here.

Eupodiscus.—Cells circular; eleven transparent oval cells near   PLATE 2.
margin ; single row of vertical cells, with spinous margin, like a pro-   Fig. 1.
jecting fringe of silex.
There were also found a great number of smaller forms like B in   2000 fathoms
this sounding—probably Thalassicolla.

Coscinodiscus.—Cells circular; margin smooth; colour pale blue in   Fig. 2.
centre, margin brown ; upper valve convex at centre.

Lagena Vulgaris.—Of ovate form; neck long and contracted; orifice   PLATE 3.
perforated ; tube two-thirds ; transparent and hyaline texture ;   Fig. 1.
foraminæ not visible. This is a variety of the form of Lagena Vulgaris.

Asterolampra (Moorheadii).—Centre oval, surrounded by one row   Fig. 2.
of cells ; equally formed except the median ray, which swells on each
side of the oval ; the marginal semicircular discs are covered with
minute areolæ gradually decreasing in size to the margin.*

* This pattern was also transferred to ivory, and appeared (if possible) more beautiful
than the Asterolampra Bishopii. It was a nice way of perpetuating these rare and
pretty forms. Diameter, $2\frac{1}{4}$ inches. This I named after my friend Dr. Moorhead, of
Weymouth.

PLATE 3.   *Foraminiferæ.*

Fig. 3.   ·   *Entosolenia* (Marginata), or properly Lagena Vulgaris, Substriata
Marginata.—Shell a perfect oval, with tubular neck, swelling midway ;
silicious margin, fluted and radiating ; surface marked with longitudinal striæ.   × 1,200 diameters.
Side view A.   Orifice at base, varies from all Lagena forms by
silicious margin.

Fig. 1.   *Entosolenia* (Globosa).—Oval form, extremity having a small mucro ;
tube not visible ; surface striated, with perforations, possibly foramina ;
aperture at extremity.   Very rare.

Fig. A.   *Bulimina* (Elegantissima).—Oblong and spiral; bluish white colour;
septalplane oblong, anterior broadest; perforations just visible.

Fig. 2.   *Nodosaria* (Radicula, fractured).—One segment, probably primordial
one with neck ; external longitudinal costæ visibly marked.

19

Foraminiferæ.   PLATE 4.

Entosolenia (Globosa) fracture (?)—Tube straight, and sexagonal Fig. 1.
form; aperture above the form of the shell, covered with irregular
striæ; hyaline texture.

Miliolina (Seminulum).—Five segments; septal aperture hidden in Fig. 2.
second segment; foramina on outer segment distinct.

Miliolina (Disciformis).—Segments irregular, outer one marked Fig. 3.
with longitudinal lines, and foramina pittings; septal aperture large
and oblong.

Textularia (Variabilis).—Segments arcuate—twenty in number; Fig. 4.
minute foramina absent from some portions; orifice distinct.

Polymorphina (Acuminata).—Variable number of oblong segments, Fig. 5.
sides unequal; segments prominent, seen in the centre; texture hyaline
and glossy; orifice not seen; perforated by minute foramina; silicious
mucro or spine projecting upwards.

Bulimina.—Segments arranged in pairs; form compressed; brown Fig. 6.
rough exterior; convolute; foramina perforations minute yet distinct.

o 2

PLATE 5.   *Diatomaceæ.*

Fig. 1.   *Nitzschia* (Angularis).—Valve lanceolate puncta in a single row on margin ; longitudinally striated ; size about .005 inch; centre light brown, indistinct.

Fig. 2.   *Triceratium.*—(Front view.)

Fig. 3.   *Synedra* (Fulgens).—Inflate at centre, and at extremes ; striæ only on each margin ; numerous. .0065 inch.

Fig. 4.   *Synedra* (Tabulata).—Taper towards extremities ; striæ marginal : numerous.

Fig. 5.   *Synedra* (Radians).—Extremes obtuse ; striæ across ; numerous.

Fig. 6.   *Navicula.*

Fig. 1.   *Tryblionella* (Hopeii, named after Admiral Sir J. Hope, G.C.B.).— Oval, =.002 inch ; striated rows of circular cells ; no central nodule perceptible with ⅛-power.

Fig. 2.
Fig. 3.   *Cymbella.*—Extremities obtuse ; striæ indistinct.   (Three views.)
Fig. 4.

Fig. 5.   *Diatom*, pin-headed.

Fig. 6.   Octagonal, fractured (?) Diatom (?) Striæ across, probably *Licmophora.*

21

Diatomaceæ—(continued).   PLATE 5.

Melosireæ (?)—Nearly circular. Two drawings are here given—one Figs. 1 & 2.
with ¼-power, the other ⅛, =540 and 900 diameters. When touched
under the glass the form fell into detached rings, as shewn in A.

Aulacodiscus.—Very beautiful ; but as the nodules were indistinct Fig. 3.
it may be found a Cyclotella ; cells not reaching to margin ; processes
two, but indistinct ; cells circular, blue and pink hue in centre ; eleven
dark cells, with alternate transparent ones near margin ; central nodule(?)
A fracture shown near centre.

Views 1 and 2.   Fig. 4.

PLATE 6.   *Polycystinæ.*

FIG. 1.   *Epithemia* (Rupestris (?))—Arctuate slightly ; striæ very distinct ;
eleven bands, pink hue.

FIG. 2.   *Vorticella.*—Hyaline texture ; greenish colour.

FIGS. 1 to 13. These minute silicious skeletons, of various forms and structures,
abounding where those of the Foraminiferæ and Diatomaceæ are usually
found, but in almost every instance, from their very delicate nature,
are fractured, are here pictured to shew their beauty and variety of
pattern, which are for the most part of a glassy transparency, except
in a few cases where they have become irregularly coated with carbonate of lime.

Many delicate glassy frameworks of fantastic and beautiful shapes
were always mixed with the material, and in various hexagonal, sexagonal, and spherical forms.

These few sketches will shew the beautiful and fanciful arrangements
of the Polycystina, among which are most conspicuous :—

> Radiolaria,
> Podocyrtis Mitra,
> Eucyrtidium Elegans,
> Astromma,
> Lychnocanium Falciferum,
>   Ditto   Lucerna,
> Eucyrtidium Tubulus,
> Rhabdolithus Sceptrum,
> Haliomma.

FIG. 11.   Fragment of *Dentalina* (sub.)—Two segments ; costate ; curves shell
shaped ; opaque.

23

PLATE 7.

Euodia (Chimmoenses).—O'Meara.

Fig. 1 shews the natural form under a small power. Fig. 1.

Fig. 2. Fig. 2.

This form differs in outline and sculpture from those described by Fig. 2 A.
Ralfs, in " Pritchard." It differs from E. Brightwellii, in which the
lower margin is concave. Compared with E. Gibba it is broader,
and the outer margin semi-circular. Towards the extremities the
dorsal margin bends inwards towards the ventral. Several examples
all presented the same outline.

The sculpture of the valve is punctate, and the puncta very dense,
and, towards either extremity, arranged in parallel lines across the
valve. No appearance of a nodule in the ventral margin.

Different views of the Euodia. Figs. 1 to 4.

PLATE 8.   *Foraminiferæ.*

Fig. 1. .   *Crystellaria.*—Succession of chambers; shell unequal sided, but completely formed—the last formed whorl scarcely visible. These shells
were in great numbers and sizes, and very perfect. This shell is
evidently of vitreous formation, having a glassy transparency, and the
perforations distinctly seen on the surface.

Fig. 1.   *Nonionina.*—This is an equal-sided shell, resembling Nautilus
Spirula, of twelve chambers, very beautifully and regularly formed.

Fig. 1.   *Coccospheres,* of one, four, and nine chambers, quite transparent, and
Fig. 2.   appear like thin discs of glass, of unequal diameters, placed one on the
Fig. 3.   other—the margin of one being seen quite distinctly through the other.

25

Foraminiferæ.

Rotalia.—One species alone is here figured in detail externally ; it is generally characteristic of all that numerous and widely-diffused group.

Fig. 1.
2000 fathoms
Lat: 47° 3′ N.
Long. 23° 21′ W.

It will be here observed, that all the chambers, eleven in number, are visible. (There are frequently thirteen chambers).

The perforations for the foramina are very distinct, and the indentations through the tumuli easily seen. The newly or last formed chamber is here transparent and colourless, while those near the first formed chamber at the spire are coated with a light brown calcareous deposit. All these forms were, by the high temperature of the Gulf Stream, 42° to 71°, in which they were found, more fully developed than those a few miles north in Polar waters, being in many instances double the size of them.

(*Polythalamus.*)

Globigerina.—In this four-chambered form is seen the aperture or orifice which is carried from chamber to chamber, probably assisting the Sarcode body to convey material to the inner chambers, or possibly for some other useful purpose as yet unknown to naturalists. <small>Fig. 2.</small>

(*Monothalamus.*)

Orbulina.—A simple calcareous spherical shell, found in great abundance, being a single cell or chamber, with its wall irregularly perforated (well known and described). In none of these forms (which were carefully examined) could any aperture be found ; probably the little inhabitant had enveloped and imprisoned itself, not intending to add to its size or form by the multiplication of other chambers, but living a perfect independent life ; and it is therefore <small>Fig. 3.</small>

D

PLATE 9.   difficult to conceive how it can be the detached reproductive segment of the Globigerinæ, whose cells, or chambers, are not in any instance spherical.

FIG. 3.   In one or two of these Orbulinæ, out of twenty which were examined, an orifice—no doubt a fracture—of irregular form was noticed, but void of that rounded, transparent, smooth finish which was found in all other specimens of the Globigerinæ.

Some of the perforations (irregular in size, form, and relative position) went completely through the wall for the purpose of extending the pseudopodia; others indenting only a portion of the wall. The external parts of the perforations were rounded and smoothed off with a beautiful transparent finish.

FIG. 4.
A. side.
B. end.   Three views of a new form of Foraminiferæ. Interior filled with minute calcareous formations; shape ovate; orifice white; body brown, and encrusted with particles of sand embedded in the shell; striated; margin white.

FIG. 5.   These few forms are of course well known and described; they are figured here simply to lead to those other forms of Foraminiferæ which are either not so well known, or are new to the general observer.

FIG. 6.   *Dentalina.*—Shell smooth; straight; five oblique segments; transparent mucro; hyaline texture; centre of each segment transparent; first and second chamber partially coated with carbonate of lime.

Foraminiferæ.   PLATE 10.

Lagena-form—(Monostegia).   — -   Fig. 1.

Triloculina.—Many varieties of this curious and simple form were 2000 fathoms
found in this sounding. Some of the Lagenidæ were replete with Lat. 47° 3' N.
Long. 23° 21'
novelty and interest. W.
A front and back view of this form is here given (Spiroloculina (?) or
Triloculina,—Miliolina), although differing in some points. It was
porcelainous, of a beautiful opaque white; when broken under the
lens (with difficulty) the orifice was full of minute granules, and the
chambers, or segments, were found irregular; there were no appear-
ances of pitted indentations on the surface. Back view, A: An
enlarged drawing of the orifice is given here; oblong; tooth simple;
texture porcelainous; opaque and glossy.

Polymorphina.—In this genus the chambers are composed of a Fig. 2.
double scries, glassy and transparent, interlaid one with the other,
having a dark band on one side of each cell; this shows a curious
alternation of the chambers, and may be considered one of the
Lagenidæ; texture smooth and glossy; foramina indistinct.

Biloculina (Ringens).—Oval; septal orifice round; tooth not seen; Fig. *a.*
porcelainous; opaque and glossy.

Polymorphina (Acuminata).—Oblong; several elongated segments; Fig. A.
sides differing.

Entosolenia (Marginata).—Elongated shell; smooth and transparent; Fig. 1.
narrow margin; none at lower extremity; foramina perforations seen;
septal orifice communicating with internal tube, passing into the

D 2

PLATE 10. cavity of the shell, twisted at the lower extremity in a hook-curve.
A rare instance.

Fig. 2.   *Entosolenia* (Quadrata).—This form, with tube projecting down-
wards internally half way, so much resembles the Actinia Crassicornis
of the Filliferous Capsule of the Polypii, that I refrain from entering
into a description of it until better known. It was not subjected to
a high power. A parallelogram with rounded extremities.

Fig. 3.   *Lagena Vulgaris.*—Semistrata and punctata ; costa at the tapering
neck only ; striæ represented at base by circles of minute tubercules.
This well-known form, with its long tubular neck and inverted
lip, except that the contracted portion of this was striated, while
the body was covered with rough tubercules, similar to some forms of
Foraminiferæ, of which it seems one. It curiously happened, while
Fig. 4.   examining this, a second appeared (the *Nodosaria*) plainly showing
(although fractured) the course of additional chambers. In this the
tube was quite visible, but not the inverted lip ; the pseudopodial
canals and tubercules were also indistinct.

Fig. 4.   *Lagena Vulgaris* (Var Perlucida).—Costa not distinct ; form ovate ;
mucro at posterior extremity.

Fig. 1.   *Miliolina.*

Fig. 2, 3.   Two separate forms of *Spiroloculina.*

Fig. 4.
Fig. a.   Front and side view.

29

Foraminiferæ.   PLATE II.

Polycystinæ.

Acanthometrina.—Spinous, with sexagonal cells, beautifully and   Figs. 1 & 2.
regularly arranged.

Lithocyclia (Ocellus).   Figs. 1, 2, 3.

Frustrella (Concentrica).

                                                   Fig. 1.

                                                   Fig. 2.

PLATE 12.   *Polycystinæ.*

Fig. 1.   Astromma.

Fig. 2.

Fig. 3.

Fig. 4.

Fig. 5.

Fig. 6.   Diatom (?)—Two of these forms found.

Fig. 7.   Glassy tube-like, probably foramina.

Fig. 1.   Polycystina of various forms.

Fig. 2.

Fig. 3.

Fig. 4.

Fig. 5.

Fig. 6.

Fig. 7.

Fig. 8.

Fig. 9.

Polycystinæ.

PLATE 13.

Lychnocanium (Falciferum).—Fracture.

Fig. 1.

Fig. 2.

Eucyrtidium (Tubulus).

Fig. 3.

Numerous glassy frameworks.

Fig. 4.

SECOND LINE
Fig. 1.

Numerous forms.

Fig. 2.

Lychnocanium.

Fig. 3.

Eucyrtidium Elegans.

THIRD LINE
Fig. 1.

Podocyrtis Mitra.

Fig. 2.

Lychnocanium Lucerna.

Fig. 3.

Many of these silicious forms possess extraordinary beauty and variety. All were probably dead forms, as in no instance was even the remains of the Sarcode body observed.

PLATE 14.

Sponge Spiculæ.

It will be seen that in this small bunch of Sponge Spicules the varieties of forms are almost endless ; some are straight, some curved, others pin-headed and studded with small spikes at regular intervals ; some pointed at both ends—some at one, resembling a needle. The most delicate and beautiful are those palm-shaped and bifurcate, curving backwards and upwards, hook-shaped and barbed ; some are short and stunted at one end. Indeed, an endless variety !

In many of these will be noticed an internal canal (much resembling a thermometer tube). They are evidently silicious deposits, but I think their origin is still attended with doubt.

These spiculæ are from the east coast of Newfoundland, at a depth of 3,900 feet.

1. ASTEROLAMPRA
(*Bishopii*)

Fig. 1.

MILIOLINA BICORNIS
(FORAMINIFERÆ.)

2. TRICERATIUM
(probably *T. Nebulosum*)

Fig. 3.

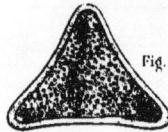

Fig. 2.

FORAMINIFERÆ.

1. MILIOLINA - *Bicornis*
2. SPIRILINI
3.

Fig. 1.

Fig. 2.

Fig. 3.

4.

DIATOMACEÆ

Plate II.

1. COSCINODISCUS

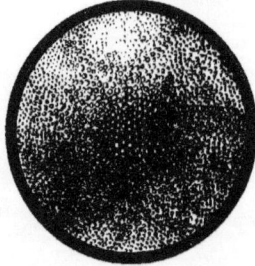

Fig. 1.

2. ACTINOPTYCHUS
3. COSCINODISCUS

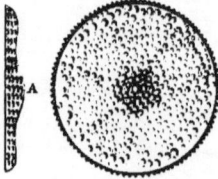

A   Fig. 2.

Fig. 3.

1. EUPODISCUS.
2. COSCINODISCUS.

Fig. 1.

A

Fig. 2.

B *Many of these, very minute, on same field.*
(THALASSICOLLA.)

Malby & Sons, Lith.

1 LAGENA VULGARIS
2 ASTEROLAMPRA (*Moorheadii.*)
3 ENTOSOLENIA - *Marginata.*

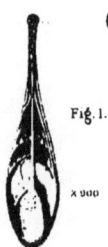

Fig. 1.

× 900

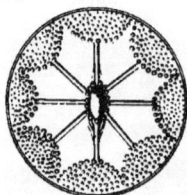

Fig. 2.

FORAMINIFERÆ.

Front

Side

Fig. 3.

A

× 1200

Orifice?

1 ENTOSOLENIA - *Globosa.*
A BULIMINA - *Elegantissima*
2 NODOSARIA *Radicula.*

Fig. 1.

A

Fig. 2.

1. ENTOSOLENIA — ... ...
2. SPIROLOCULINA   } Miliolina
3. SPIROLOCULINA

Fig. 1.

Fig. 2.

Fig. 3.

TEXTULARIA — *Variabilis.*
Fig. 4.

POLYMORPHINA — *Communis.*
...LINA — *Princeps.*

Fig. 5.

Fig. 6.

1. NITZSCHIA
2. TRICERATIUM *(probably front view.)*
3. SYNEDRA
4. SYNEDRA
5. SYNEDRA
6. NAVICULA

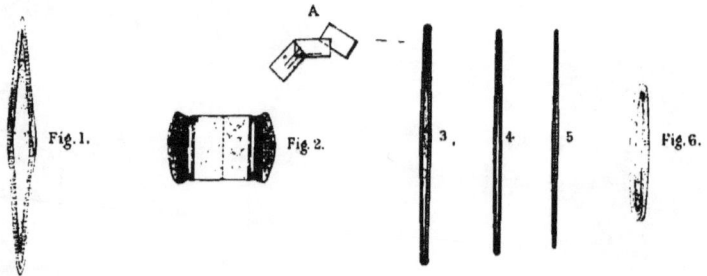

A

Fig.1.   Fig.2.   3.   4.   5.   Fig.6.

1. TRYBLIONELLA   *(Hopeii..)*
2.3.4. CYMBELLA
5
6   LICMOPHORA ?

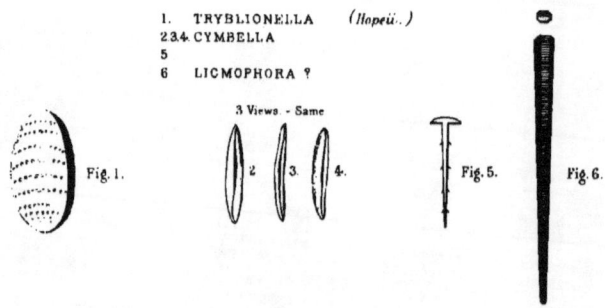

3 Views - Same

Fig.1.   2.   3.   4.   Fig.5.   Fig.6.

1. 2.   MELOSIREA.
3   AULACODISCUS
4.

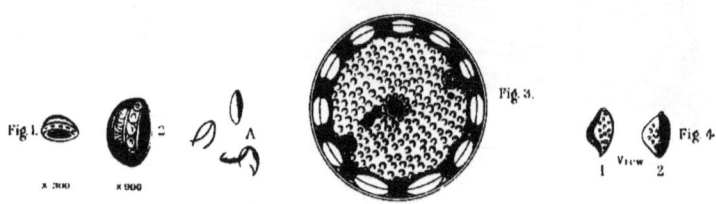

Fig.1.   Fig 3.   Fig 4.

Fig.1.   × 300   × 900   1   View   2

EPITHEMIA. *(Rupestris.)*

Fig. 1.

Fig. 2.

VORTICELLA.
Hyaline texture
Greenish hue

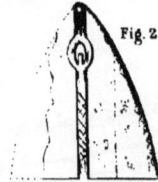

A few forms of Polycystina *and* Diatomaceæ
from various depths.

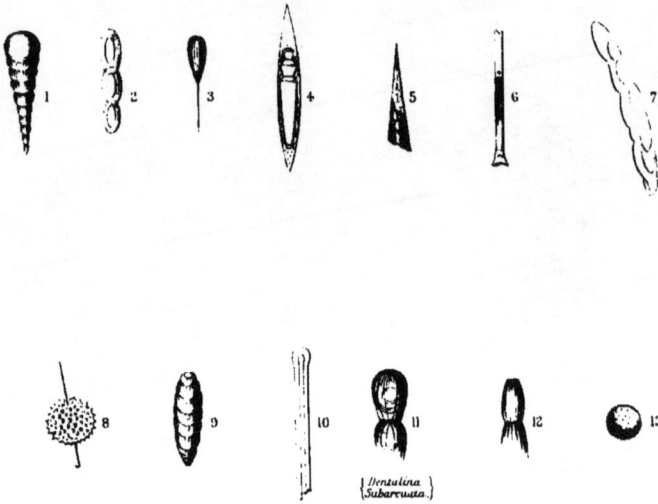

{ *Dentalina*
Subarcuata. }

EUODIA. *Chimmo-ensis (O'Mearn.)*

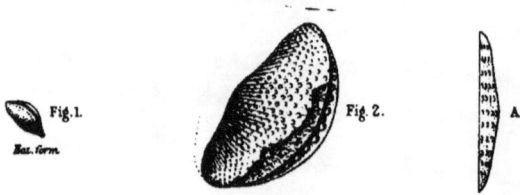

Fig.1.

Bas. form.

Fig. 2.

A

EUODIA.

Fig.1.

Fig.2.

A

Section.

A

Fig.1.

2

3

4

Five positions of Euodia.

W. C. fecit.

Malby & Sons, Lith.

CRISTELLARIA

Fig 1

Fig 2

NONIONINA

Fig 3

Fig 4

Fig 5

COCCOSPHERES

Fig 1

Fig 2

Fig 3

DENTALINA *legumen.*

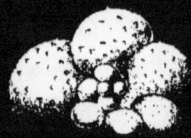  

Fig. 1. Fig. 2. Fig. 3.

Fig. 4.

DENTALINA

Fig. 5.

Fig. 6.

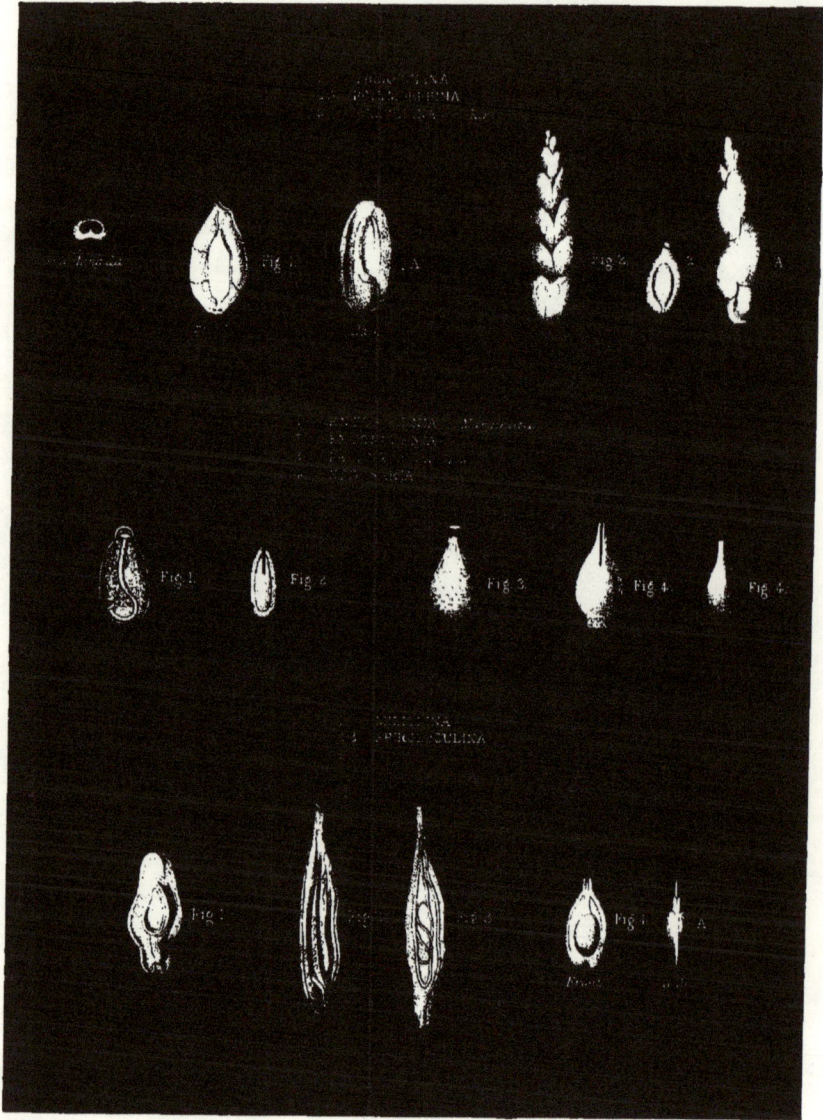

1. ACANTHOMETRA
2. HALIOMMA — *Heteracantha*

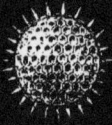
Fig 1.

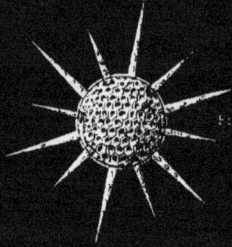
Fig 2.

Fig 1.

Fig 2.

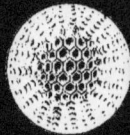
Fig 3.

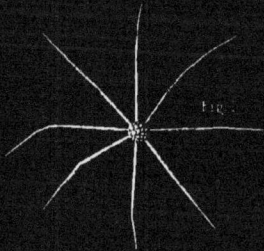
Fig 1.

Fig 2.

ASTROMMA.

Fig 1.   Fig 2.   Fig 3.   Fig 4.

Fig 5.      Fig 6.   Fig 7.

Polycystinæ

LYCHNOCANIUM — *subsferens*

EUCYRT/DIUM — *succinus*

Fig 1   Fig 2   Fig 3   Fig 4

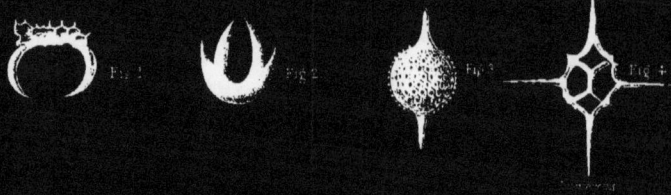

1
2
3   LYCHNOCANIUM
4

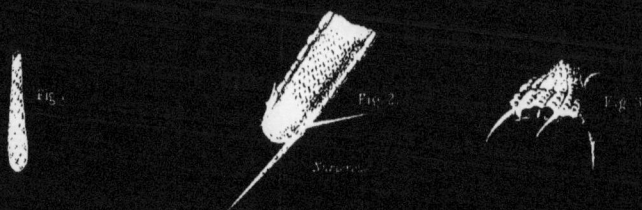

Fig 1   Fig 2   Fig 3

Nucleus

EUCYRTIDIUM ELEGANS
PODOCYRTIS MITRA
LYCHNOCANIUM LUCERNA

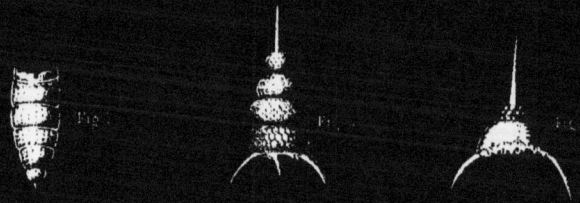

Fig   Fig   Fig

SPONGE SPICULÆ.

W. C. fecit.

Malby & Sons, Lith.

Note.—It will be seen that some of the forms on these Plates are not named. It is difficult with only a general view to do so correctly, as there was neither time nor opportunity to go into every detail. I have preferred leaving them thus, rather than calling them by inappropriate or incorrect names. But I believe, that when compared carefully with all previously discovered forms, both recent and fossil, many will be found new.

E

APPENDIX.

Soundings and Temperatures in the Gulf Stream. By Commander W. Chimmo, R.N.,
F.R.G.S., F.R.A.S., &c.

Towards the latter part of the year 1868, after H.M.S. *Gannet* had been upwards of three years on the North American and West India Station, she was ordered by the Admiralty during her homeward voyage to define the northern limits of the Gulf Stream, and to take deep soundings and temperatures within those limits.

Sailing from Halifax, in Nova Scotia, on the 1st of July, the ship passed from water whose surface temperature was 51°, to that of 61°, in less than an hour—shortly afterwards to 64°; showing that the Gulf Stream water had been reached since leaving that place.

Lat. 43° 20′ N. ; long. 60° W.—South of Sable Island, 30 miles, a sounding was obtained of 2600 fathoms, or 15,600 feet—nearly 3 miles; with a weight of 232 lbs., and the ingenious machine invented by Brooke, the rod brought up, after four hours' patient hauling, Foraminiferæ in various forms, chiefly Globigerinæ clusters of three, four, and five chambers. The interior of those fully developed was coated with an apparently fine crystallized, many-coloured, quartzose sand : of these forms some were circular—flat and plate-shaped, having a smooth interior rim (the Polycystina) : the outer margin serrated, and the centre coated with the same granular particles. Others hemispherical, some single and globular ; others, fragments thin and transparent as water. Intermixed with these were minute particles of transparent many-coloured crystals, with coccospheres in all stages of growth and size.

The towing-net collected seven species of Crustacea, one Cornucopia, and a Janthina fragilis ; the dye from which latter, when placed in a wine-glass of clear water, coloured the whole a rich mauve. A very small portion of this apparently inpalpable adhesive mud, when diluted, and placed under the microscope, showed a field of the most perfectly-formed organisms.

The ship next sailed to the western edge of the Grand Banks of Newfoundland, where a sounding of 1500 fathoms brought up what appeared, under a common glass, minute particles of transparent quartzose sand, with globular forms of calcareous formation ; also some algæ with parasitical attachments, probably of lime, but all formed by animal life from carbonate of lime from ocean waters.

The temperature of this mud or "Ooze," as it will be called, was 56° ; but at a depth of 1000 fathoms the thermometer showed 40°·3, and at 500 fathoms only 39°·5, so that the mud probably changed its temperature in passing through a stratum of warmer water, as the sea-surface was 60°. This showed an under stratum of very cold water ; there being a difference of 20° between the surface and 500 fathoms, and possibly so at a very much less depth.

Having run north of the limit of the Gulf Stream, again stood to the southward, and soon came into warmer water, at a temperature of 60° ; from a cold, damp, penetrating fog, into a mild and summer-like atmosphere ; 1500 fathoms was again found, and the cup brought up the usual grey impalpable mud (ooze). The towing-net collected a beautiful float of the Nautilus, having 13 chambers, and a fragmentary valve of a delicate fluted Pecten.

The temperatures were precisely the same as in the former sounding, except that the surface was 65°, and at 100 fathoms the thermometer showed 50° ; a difference of 15° in only 100 fathoms—another proof of the Gulf Stream being merely superficial.

At day-dawn this morning, to the great surprise of every one, we saw an old Labrador friend—a huge iceberg—having a warm bath in a temperature of 62°, double that of its own. Although it was still 150 feet high, and nearly 400 immersed, it was quickly and perceptibly undermining, decomposing, splitting with loud reports, and floating away in large portions with the easterly current.

It curiously happened that this immense iceberg stood in the very spot—30 miles south of the edge of the Grand Bank—where not only the deepest waters of the Atlantic were supposed to be, but where we intended to get a sounding to ascertain if this were the fact : the result showed it was not so.

Sail was furled, steam got up, and the Gannet ranged up as near as was prudent under the lee of our chilly friend ; and in the midst of a thunder storm, with Brooke's rod and weights, obtained at a depth of 1400 fathoms the same "Ooze," disproving the idea of the deepest water being here. This depth appears to be not only the usual one, but also the general slope of the Banks, as well as the universal character of their formation.

By the temperatures here obtained, the same stratum of cold arctic water was passing under the warmer waters of the Gulf Stream. The rod brought up a small portion of feldspar with glittering particles of mica, evidently deposited there by icebergs from Davis Strait, and that very recently.

The ship now sailed east for the spot where Lieut. Sainthill, in lat. 42° 37′ N. and long. 41° 45′ W., obtained, in 1832, 100 fathoms on sharp rocky bottom, bringing up on the arming of the lead "fine bluish ashes ; " and he was under the impression that he was over a submarine volcano in a state of eruption. At 2 P.M., on the 12th of July, this position was reached, and with a heavy weight 4300 fathoms of line ran out, and no bottom !

It was somewhat remarkable that about here, within a radius of some few miles, many indications of shoal water had been from time to time seen and reported, one having as little as 35 fathoms on it. To one of these, called the "Milne Bank," with only 80 fathoms on it, we were now steering. It had been found by H.M.S. Nile, in 1864, on her homeward-bound voyage ; and, under most favourable circumstances, soundings of 80, 90, and 100 fathoms, "fine sand and ooze" brought up.

Also, about this vicinity, the currents are found very strong, and a little further east very variable in direction ; sometimes running with a velocity of 2, 3, and even 4 miles an hour to the eastward, and in some places forming a complete "race." If neither banks nor shoal-water exist here, it is not easy to account for this additional effort of the Gulf Stream ; unless, indeed, it is the mass of water brought from the South Atlantic by the south-east trades, adding to its volume and to its velocity.

Lat. 43° 30′ N. ; long. 38° 50′ W.—At 4 P.M., on the 15th of July, we were on the 80 fathoms ! The rod and weight of 230 lbs. let go, and as each 100 fathoms ran off the reel it caused some excitement, as at each fathom it was hoped the bank would be struck. 2280 fathoms, 13,680 feet, ran out. There was no bank there. The rod brought up "ooze" abounding in animal, vegetable, and mineral remains !

E 2

Here the thermometers were sent down to ascertain specially if cold water existed at any depth. One thermometer burst at 1400 fathoms. Water was brought up from a depth of 1500 fathoms,* containing small and delicate particles of algæ of various bright colours, showing possib'y, that light had penetrated to that depth ; but there was no sign of animal life† to the naked eye.

Another sounding for the bank was tried, and 2600 fathoms obtained ; the rod bringing up from the same vast mass, countless thousands of the same character as those found a day or two previous at a nearly similar depth, except that the Globigerinæ were in clusters, and in those *fractured* there was apparently a hard, compact, crystallised, fine sand.

The fractured Globigerinæ in this sounding were very beautiful, showing marginal walls of vertical crystal formation, clear as water, the fractured globes or cells containing (apparently) minute quartzose sand. Thinner glass-like forms of air-like globules, in irregular order, were probably Coccospheres.

A small convex portion illustrated beautifully the radiating perforations or canals of the Foraminifer, both direct and diagonal ; and some few irregular particles of Diatoms flexible and multiform.

Some of these also show the horizontal layers of each wall, added layer to layer—the outer ones thickening, and the external layer becoming coated with tubercules ; the interior are of an enamel transparent smoothness.

The heat in the Gulf Stream was found at times very oppressive, and reminded us all of the climate of Trinidad in the wet season. The thermometer in the shade was 82°, in the sun 96° ; the warm vapour arising from the heated water made one feel languid, lazy, and sleepy, and was very debilitating.

By the temperatures obtained from actual observation at 300, 500, and 1000 fathoms, the waters were in all cases warmer than the corresponding depths north of the Gulf Stream. This is, of course, very natural, but it is as well to have it from actual observation ; and this would argue in favour of bodies of warm water being brought up from the coast of Africa by the south-east trades, and, accumulating with those of the Gulf Stream on the position assigned to the Milne Bank, assisting materially in adding to its velocity and irregularity.

Stood north again for Polar waters, which were soon felt by the temperature of the sea-surface changing in 2½ hours 14°—viz., from 72° to 58°,—giving again the northern limits of the Gulf Stream. The air also gave proof of this again, for in an hour we passed from a close uncomfortable heat to a chilly cold, which compelled all hands to put on warm jackets ; and, as a natural consequence of this change, soon followed a dense fog !

Ran for the Flemish Cap, on which we sounded and obtained 80 fathoms. Stones, Feldtspar, and various coloured quartz, with some few Foraminiferæ even in these shoal waters.

Sounded midway between the north part of the Flemish Cap and the Grand Bank, to ascertain if there was any connection, or if they were separated by a deep channel. 250 fathoms was obtained, showing that it *was* part of the bank, but having a rocky nucleus, about which the soil brought down by the ice accumulates ; but the Polar current over it is sufficiently strong to keep the rock

* Temperature 42°.

† The temperature of the air was 77°
  ,,   ,,   ,, sea ,, 73°
At 100 fathoms below it was 62° = 10 degrees less.
At 300 ,,   ,,   55° = 20 ,,
And at 1000 ,,   ,,   42° or 30 ,,
so that cold Polar waters were passing underneath at 200 fathoms below the surface.

bare. On two occasions it bent and turned the iron cup of the weight in 90 fathoms; here at 250 fathoms the temperature of the sea was 38°, while at the surface it was 50°; the air being the same. The south part of the Cup is not, however, united to the Banks, for 700 fathoms, and no ground was obtained between them.

On the parallel of 46° latitude, at a distance of 25 miles from the edge of the bank, sounded in 1000 fathoms, bringing up large quantities of rounded particles of quartz of various colours.

Here a section of the slope of the bank was made, showing its ascent, formation, and the nature of these vast banks. From 1000 fathoms—of coloured quartzose sand, to 650—of silicious spicules of sponges ; then to 450—green mud; 150—quartzose sand; 60—stones ; 55—stones, sand, and fish-bones ; and the latter told us that we were on the Grand Banks.

Passing over and searching for the "Jesse Ryder Shoal" of 4 fathoms, which was found *not* to exist, we put over the dredge and dropped on a perfect colony of star-fish (Ophiocoma) of all sizes, from half an inch to 3 inches in diameter.

In a very dense fog steered for St. John's, Newfoundland, where we arrived on the 24th July, to rest for a few days after the work in the Gulf Stream. It was cold, raw, and foggy ; but we were very glad to drop anchor in its snug and secure harbour, free for a while from all the cares, anxieties, and perplexities necessarily attending deep-sea sounding.

Having again prepared lines, instruments, and chronometers for a second voyage, sailed on the 27th August for the north extreme of the Gulf Stream, and which was reached two days afterwards —the sea temperature rising suddenly from 53° to 61°.

Lat. 44° 3′ N. ; long. 48° 7′ W.—Here soundings were again obtained with rod and heavy detaching weights in 1650 fathoms, bringing up Foraminiferæ in all stages, whole and fragmentary, having from two to six cells or chambers in clusters, spherical, plate and flat-shaped Polycystina, with a few spicules of sponges, as well as coccoliths.

Temperatures of under strata of currents were obtained, again showing that at 1000 fathoms the water was 39° 5′, and at only 50 fathoms below the surface (which was 61°) it was 43°, or 18° colder !—air being 61° ; another proof of the bare superficial Gulf Stream.

Another cast of the lead on the supposed position of the Sainthill volcano quite disproved the existence of this vigia within a radius of many miles.

We were approaching for the second time the "Milne Bank," of 80 fathoms ; and although 2300 fathoms was obtained on it a short time since, yet there was still a hope that the second attempt would be more successful, particularly as a telegram had reached me from England to the effect that "there was no doubt of the Milne Bank, as bottom was brought up *three* times ;" and indeed so it would appear. But on the 3rd of September (lat. 43° 40′ N., long. 38° 50′ W.) the lead was again let go and 2700 fathoms obtained, the rod bringing up a small but precious portion of Foraminiferæ.

The towing-net gave another rich haul of Hyalæa—Atlanta and Spirula—with three specimens of Nautilus cornucopia (I believe the latter to be Operculate).

It is interesting to find how the different species of these delicate ocean-shells have their own special haunts and feeding-grounds. In one place the Atlanta are taken in numbers, with scarcely any others ; in another a net full of Hyalæa tridentata ; then quantities of Spinosa or Radiata ; lastly, a bag of Jauthina fragilis ; but these latter are more generally distributed than others. All these delicate creatures are found more numerous on the surface at the sun's rising and setting, and very abundant during light showers of rain.

Near the supposed position of this bank we sounded at short distances for some days with more than a thousand fathoms of line ; but in no case was there any indication of this bank. The last effort to sound in 1000 fathoms north of its vicinity will not easily be forgotten ; it was obtained under many and great difficulties. The sea had risen to a fearful height in a very short time, which threatened to roll all the boats from the davits. My steam-cutter *Torch* did start. There was scarcely any standing on the deck. All the thirty-five men on the starboard side, while hauling the line in, lay down like dominoes, or rather were thrown down on the line.

Lat. 43° 30′ N. ; long. 38° 5′ W.—Sounded again with heavy weights in 2000 fathoms, bringing up Foraminiferæ in various stages of growth ; and what gave interest and value to this sounding, was the fact that icebergs had reached these eastern limits, proved by the presence of a piece of stone (Feldtspar) three-quarters of an inch in size, deposited undoubtedly there by a berg, and brought up in the valve.

Lat. 43° 43′ N. ; long. 37° 47′ W.—On the 5th of September a sounding was taken at 1930 fathoms ; the rod came up with its spring broken, but retaining sufficient of the bottom to show that it was down. Foraminiferæ, the usual deep-sea characteristic, appeared, mostly young and small, with transparent cells ; about 6 per cent. of all these were free from fracture, all the remainder fragments.

Before leaving the vicinity of this supposed bank, the temperatures here obtained with new and delicate thermometers at 2000 fathoms was 42°—rather a higher temperature than expected. The air was 68° ; the sea-surface 69° ; while at 100 fathoms it had fallen 10°, and at 400 20° ! At 1000 fathoms it was 43°, after which it fell but 1° in 1000 fathoms.*

Great quantities of Salpœ and Medusæ came up entangled with the line, doubtless caught in its quick descent of 500 fathoms in 3½ minutes.† Their orange-coloured stomachs, situated in the centre of the transparent gelatinous sacs, came in quite perfect on the line.

Lat. 43° 39′ N. ; long. 36° 46′ W.—On the 6th September we gave our departing and final cast of the lead in this vicinity, getting 2060 fathoms ; the rod bringing up Foraminiferæ, small stones, and some few Diatoms.

We were now leaving the regions of the Globigerinæ and Lime formations, changing them for those of Silicious deposit. Nearly all in the last sounding were Diatoms, with but few Globigerinæ. A thermometer was sent down to 2000 fathoms and proved the last temperature at the same depth, showing 42°·5.

To complete a series of 100-fathom temperatures, advantage was taken of a fine day with smooth water—both indispensable requisites in sounding for temperatures, as the smallest jerk or vibration moves the indices and the reading is destroyed, the results being only deceptive. The thermometer went down to 1500 fathoms, and in no instance did it show less than 42°·5, fully proving the high temperatures obtained on former occasions, and this would prove the entire absence of an under Polar current here ; and further, that the waters of the Gulf Stream here united with other waters, decreasing thereby its strong easterly set, which was here found on many occasions to be variable. The temperature of the surface was 71°.

From the authority of a few scattered observations, it has been asserted that the water of the ocean, at a depth of 12 feet, was of a higher temperature than at the surface. This was proved to

---

* Showing a great uniformity of temperature after the first 500 fathoms.

† Or equal 14 feet in 1 second, which equals 1 mile in 6 minutes.

be correct, although remarkable, by 45 carefully-obtained observations between Halifax and this position.

Of these 45 observations, 26 are warmer, 10 are colder, and 9 have the same temperature. The warmer are in favour of the colder, 16° in the whole, but in no one instance greater than 1°·5 ; and the greatest and most constant are noticeable to the east of the Milne Bank, after the rapid current of the Gulf Stream had been passed.

In the Pacific, off the west coast of America (the Isalcos Mountains), the temperature at 12 or 15 feet below the surface has been found to be 10° or 11° higher. This, I presume, is caused simply by excessive evaporation, as I have often found there the difference between the wet and dry-bulb hygrometer to be 9°, and on occasion 11°.

Lat. 46° N. ; long. 29° 40′ W.—9th of September, being on the position of a vigia, a very satisfactory sounding of 1650 fathoms was obtained : first disproving the existence of such a danger, and secondly bringing up the most interesting and remarkable specimen of the bottom ; showing that those minute creatures (Pteropods) which live on the surface do assist in forming the bottom of the ocean. Foraminiferæ and Diatomaceæ surrounding six dead Hyalæa shells, all perfect. These, to have been taken on the bottom, must have been dead, and for a valve the size of a shilling to have entrapped six of these, they must have been numerous indeed ; the whole area of the six was greater than the valve itself—they must, therefore, have been in such quantities as to overlap one another. Hyalæa were also taken on the surface in the towing-net ; so that here was a successful illustration that those lived on the surface and fell, after their period of existence, to the bottom.

This was a shoal-sounding compared with those around it, and silicious formations now became more numerous ; Coccospheres and other delicate forms, some resembling the Nautilus, with thirteen chambers, but devoid of the syphuncle which assists to elevate or depress the latter at pleasure, by exhausting or filling its chambers with water.

In this sounding, also, animal remains were seen, and could hardly be mistaken ; the pseudopodial foramina or radiating processes from the tubercles of the canals were regularly protruding, and at the point where the chambers intersect was a mass of minute spawn-like globules.

Inorganic fragments of some size were also seen, having a smooth concave impression, intersected with dark lines. In no instance are the shells of the Hyalæa, taken alive on the surface, so large as those found dead on the bottom ; so that it may be possibly inferred that they have died at their full growth, at the limit of their permitted existence.

A very interesting and valuable sounding was made about 180 miles E.N.E. of the last, in 1180 fathoms, showing a less depth of water by 200 fathoms than in any part of the Atlantic (not approximate to the shore). A small portion of the bottom "Oase" came up attached to a pig of ballast, which was the weight used on this occasion.

Lat. 47° 11′ N. ; long. 23° 14′ W.—On the 12th September the favourable weather, with a dead calm, induced us to sound, and a cup-lead of 112 lbs. reached the bottom at 2000 fathoms, bringing up a full cup of pale cream-colour "Oase," Infusoria, like ice-cream, and quite as cold. In this sounding were many-shaped and various-formed Globigerinæ, hemispherical and globular ; also many spheroidal organisms, in one specimen of which we counted thirteen chambers. It was from this sounding that 113 different specimens were obtained which form the subject of this pamphlet.

A fractured portion of a Globigerina cell showed that the interior wall was formed of a perpendicular transparent four-sided cell, while the exterior was perforated by narrow canals running perpendicular to the frame. The temperature at that depth was still 42′.

Our sounding now ceased, and this exciting and very interesting work finished.

It is worthy of remark that the general character of all these thirteen soundings, varying in depth from 80 to 2700 fathoms, spreading over an area of upwards of 10,000 square miles from Sable Island to the Azores, shows a remarkable uniformity both in respect of temperature and sea-bottom. One object throughout was to ascertain if in any of these organised forms animal life still existed. After arrival (as I had no microscope of sufficient power on board) they were examined for fourteen days under a powerful microscope, and in no one instance was animal life visible.

Many hundreds of the animal organisms of Foraminifore, Globigerinæ, Coccoliths, &c., with which the soft light brown and yellow mud abounded, were, after being diluted with clear water, separated from the muddy particles and broken under the lens with a finely-pointed penknife. It required some force to break them, and the sharp shock and cracking was plainly perceptible; in no instance was life visible.

The mud, when dry, is either of a pale yellow marl, light brown, or greenish brown colour; the former containing chiefly Globigerinæ or calcareous formations, the second silicious or Diatomacere, and the last silicious spicules of sponges. All are apparently soft mud until rubbed between the fingers, when gritty particles are detected. These are the Globigerinæ in great variety of shapes and numbers, some being formed in concentric layers round a transparent centre.

In the deepest waters and most distant from land were the greatest numbers of perfect specimens of the Globigerinæ found; and as the water decreased in depth and neared irregularities, so they became fragmentary. These facts suggest that, either at the lesser depth some wave-movement, or, may be, current, fractured these delicate organisms, or that their cases were broken by mollusks or other devouring agents for the softer matter in the interior, and the shelly portions then descended to the bottom.

With many experiments in water, it was found that not only were the Globigerinæ of much more specific gravity than the water, but that they sank with a rapidity truly wonderful, and invariably with the convex side downward, and in that position (which was contrary to that in which they lived) remained so.

In passing the Ooze a second time under the microscope, some new forms were detected, which will be seen in drawings exhibited before the Society;* these are for the most part of silicious formation, some having a thin, irregular, and broken coating of lime; others as transparent as glass.

The thin membrane lining in some of the Globigerinæ were also noticed, but these could hardly be the remains of the once-living animal.

Some recent Globigerinæ, which appeared discoloured (a light red), were broken; but no minute granules were inside.

In the second examination of the Globigerinæ I felt compelled to alter my views with regard to the rounded aperture noticed (which I thought may be formed by an annelid), but which I found in every form, larger or smaller, according to age and size. In some instances the apertures were in the two last chambers, the lips of which were smooth and rounded off with a transparent glass-like finish.

* Fourteen Plates containing upwards of 252 specimens.

LIST OF SUBSCRIBERS.

	Copies.
Dr. Smith, Greenhill, Weymouth	2
Professor Haughton, LL.D., F.R.S., Dublin	1
Hildebrand Ramsden, Esq., Leadenhall-street, London	1
Admiral Sir J. Hope, G.C.B., Portsmouth	1
Dr. J. Moorhead, Weymouth	2
Adderley W. Chimmo, Esq., R.N., H.M.S. "Boscawen"	1
James Watson, Esq., Hamilton, Canada (W.)	1
C. Capel, Esq., 136, Leadenhall-street, London	1
Admiral Sir W. H. Hall, F.R.S., Senior U.S. Club, London	1
William McKinnon, Esq., Balnakiel, near Glasgow	1
Capt. Clavering, R.N., Weymouth	1
John Turnbull, Esq., 1st Royals, N. & M. Club, London	1
George Bishop, Esq., Observatory, Twickenham	2
Flag-Lieutenant W. J. L. Wharton, R.N., Portsmouth	1
John Alexander, Esq., 48, Porchester-terrace, London	1
Capt. Montague Thomas, R.N., Weymouth	1
Rev. Eugene O'Meara, Hazlehatch, Dublin	1
Admiral Sir H. Kellett, K.C.B., China	1
James Chimmo, Esq., R.N., Cork, Ireland	1
John Barrow, Esq., F.R.S., 17, Hanover-terrace, London	1
Professor A. Geikie, F.R.S., Geographical Survey, Edinburgh	1
Thomas Watson, Esq., W.S., Gédinne, Belgium	1
Lieut. G. Purdon, R.N., F.R.G.S., Binfield, Berks	1
James Hall, Esq., Woodside-crescent, Glasgow	1
Captain A. B. Becher, R.N., F.R.A.S., Dorset-square, London	1
Mrs. Gregson, Toxteth Park, Liverpool	2
Robert Jamieson, Esq., W.S., Woodside-crescent, Glasgow	1
William Thompson, Esq., Weymouth	1
John Napier, Esq., Milliken, Renfrewshire	2
Miss Jamieson, Woodside-crescent, Glasgow	1
Admiral Jervis, R.N., Weymouth (Club)	1
Capt. Barrington Browne, Weymouth (Club)	1
Archibald Galbraith, Esq., Johnstone Castle, Renfrewshire	1
Professor Allen Thomson, F.R.S., Glasgow	1
J. Burrows, Esq., Grove-road, Wanstead, Essex	1
H. Crouch, Esq., 51, London-wall	1
Sir J. Matheson, Bart., M.P., Stornoway, Scotland	1
J. S. Cousens, Esq., Grove-road, Wanstead, Essex	1
John Burns, Esq., F.R.G.S., Castle Wemyss, Renfrewshire	2
T. Curties, Esq., 244, High Holborn, London	1
J. G. Tatem, Esq., Russell-road, Reading	1
H. H. Browne, Esq., F.R.G.S., Conservative Club	1
Robert Napier, M.P., Shandon, Dumbartonshire	2
Andrew Galbraith, Esq., Johnstone Castle, by Glasgow	1

	Copies.
Mrs. J. Graham, Drums, Renfrewshire	1
Hugh Cowan, Esq., Belmont, Paisley, Scotland	1
John T. Tullet, Esq., 200, Camden-road	1
H. F. Austin, Esq., Weymouth Club	1
Rev. J. Watson, Oxford	1
Rev. J. A. Ashworth, Didcot Rectory	1
Admiral Sir Rodney Mundy, K.C.B., Royal Naval Club, London	1
J. Grahame, Esq., Gordon Lodge, Reading	1
Mrs. Mackichan, 46, Moray-place, Edinburgh	2
Professor George Rolleston, M.A., F.L.S., F.R.S., &c., Oxford	2
Captain Medlycott, R.N., Ven. Sherbourne	1
R. G. Hancock, Esq., Weymouth	1
M. H. Devenish, Esq., Weymouth	1
James C. Burns, Esq., 1 Park-gardens, Glasgow	2
Anthony Perrier, Esq., Fotabeg, Cork	1
James Keane, Esq., Afghanistan House, Cork	1
J. Martin, Esq., 5, Victoria-terrace, Weymouth	1
Mrs. James Scott, Woodside-place, Glasgow	1
George Sherlock, Esq., Carrigbude, Blackrock, Cork	1
Mrs. Richardson, Ralston, Renfrewshire	2
Robert Blackwell, Esq., Carrigaline, Cork	1
Richard Griffith, Esq., Millicent, Glanmire, Cork	1
W. J. Knight, B.A., Cork	1
J. C. Bolton, Esq., Glasgow	1
Robert Ker, Esq., "	1
William Ker, Esq., "	1
Anthony Hannay, Esq., "	1
Robert Aitken, Esq., "	1
James N. Fleming, Esq., "	1
William Anderson, Esq., "	1
William Connal, Esq., "	1
Thomas Muir, Esq., "	1
Alexander Drew, Esq., "	1
John Moffat, Esq., "	1
Dr. Roberton, "	1
James King, Esq., "	1
John McEwen, Esq., "	1
William McEwen, Esq., "	1
James Morrison, Esq., "	1
Thomas Thomson, Esq., "	1
William Johnston, Esq., "	1
George Brown, Esq., "	1
John Brown, Esq., "	1
David Hutchison, Esq., "	1
John Ross, Junr., Esq., "	1
George Munsie, Esq., "	1
John Tennant, Esq., "	1
Alexander Dennistoun, Esq., "	1
Robert Dalglish, Esq., M.P., "	1
Charles Gairdner, Esq., "	1
James Hannan, Esq., "	1
James Reid, Esq., "	1
James McGregor, Esq., "	1
Rev. F. St. George, Cork, Ireland	1

www.ingramcontent.com/pod-product-compliance
Lightning Source LLC
Chambersburg PA
CBHW021626270326
41931CB00008B/890